www.heinemann.co.uk/library

Visit our website to find out more information about **Heinemann Library** books.

To order:

☎ Phone ++44 (0)1865 888066

🖹 Send a fax to ++44 (0)1865 314091

💻 Visit the Heinemann Bookshop at www.heinemann.co.uk/library to browse our catalogue and order online.

First published in Great Britain by Heinemann Library, Halley Court, Jordan Hill, Oxford OX2 8EJ, a division of Reed Educational and Professional Publishing Ltd. Heinemann is a registered trademark of Reed Educational & Professional Publishing Ltd.

OXFORD MELBOURNE AUCKLAND JOHANNESBURG BLANTYRE GABORONE IBADAN PORTSMOUTH NH (USA) CHICAGO

Designed by Paul Davies and Associates
Illustrations by Wooden Ark
Originated by Ambassador Litho Ltd.
Printed by Wing King Tong in Hong Kong.

06 05 04 03 02
10 9 8 7 6 5 4 3 2 1

ISBN 0 431 14709 4

British Library Cataloguing in Publication Data

Snedden, Robert
 Cell division and genetics. – (Cells and life)
 1.Cell division – Juvenile literature 2.Genetics – Juvenile
 literature
 I.Title
 571.8'44

Acknowledgements

The Publishers would like to thank the following for permission to reproduce photographs:
Hulton Archive pg 39; Image Bank: pg 40; OSF: /D J Cox pg 43, /A Ramage pg 17, /H Reinhard pg 21, /A Shay pg 16, /R Winslow /Animals, Animals pg 25, /F Whitehead /Animals Animals pg 24; Photo disc: pgs 5 and 22; SPL: pg 20, /Biophotos Associates pg 38/, J Burgess pg 42, /CNRI pg 14, /Eye of Science pgs 11, 32 and 33a, /K Edward pg 31, /N Kedersha/ULCA pg 7, /A Khodjakov pg 10, /K Lounatmaa pg 23, /P.Motta and T. Naguro pg 33b, /G. Murti pg 4, /Y Nikas pg 18, /D Parker pg 29, /A Pasieka pg 26, /D Scharf pg 13, /G Schatten pg 12, /S Stamers pg 6; Stone: pg 36; A. Syred pg 8.

Cover photo reproduced with permission of Science Photo Library/Dr Tony Brain.

Our thanks to Richard Fosbery for his comments in the preparation of this book, and also to Alexandra Clayton.

Every effort has been made to contact copyright holders of any material reproduced in this book. Any omissions will be rectified in subsequent printings if notice is given to the Publisher.

Contents

*Words in bold, **like this,** are in the Glossary.*

1 Introduction

Why does a pig always give birth to piglets, not to kittens? Why does a caterpillar, not a tadpole hatch from a butterfly **egg**? These might seem like odd questions to ask, but they are at the heart of the study of **genetics**. Genetics is the study of heredity, or how the characteristics of living things are passed on from one generation to the next.

All living things are made of cells. Cells are life's building blocks. A single cell alone is too small to be seen without the help of a microscope, but this tiny chemical package contains all the ingredients of life. Very simple forms of life consist of single cells. However, human beings, and all the plants and animals around us, are built from millions upon millions of cells, all working together.

Dividing cells

Every living thing, from the smallest to the largest, begins life as a single cell. This single egg cell divides and divides many times, until it grows into an adult. Cell division is how **organisms** grow. When the organism is fully grown, some cells continue to divide to replace cells that grow old or are damaged in some way.

Each tiny egg cell holds the information that enables it to grow into a complex organism such as a cat or an oak tree. A sea urchin egg cell will grow into a sea urchin, while a human egg cell will become a new person. This information is contained in **DNA**, the cell's **genetic material**. DNA is an amazing substance that acts as an instruction book, containing all the information a cell needs to reproduce itself.

This magnified photograph of a eukaryote cell shows the nucleus and other organelles. Magnification approx. x 600.

Animals like this gibbon produce offspring that are similar to themselves. It passes on its characteristics to the next generation through the genetic material in its sex cells.

Reproduction

Bacteria and other micro-organisms that consist of single cells reproduce themselves by dividing in two. Most multicellular organisms cannot reproduce in this way: they are too complex. Instead, they produce special cells from which a new organism can grow. In sexual reproduction, cells from two parents combine to form a new living thing. For sexual reproduction an organism has to make special sex cells, and this involves a different kind of cell division.

In this book we will start by looking at how cells divide, and find out more about DNA. We will then look at sexual reproduction, and the different kind of cell division that this involves. Finally we look at ways in which characteristics of living things are passed from one generation to another in their genetic material.

Prokaryotes and eukaryotes

There are basically two types of cell in the living world. The cells of bacteria, the simplest living things, have no obvious structure. In particular, the DNA in a bacterial cell is not contained in a distinct **nucleus**.

The cells of all other living things are more complex. The inside of the cell is divided up into various small compartments, called **organelles**. The organelles keep different reactions in the cell separate, and make it possible to gather all the materials for a particular job in one small area. The largest organelle in a cell is the nucleus, which contains the cell's DNA.

Bacterial cells are called **prokaryotes** (a word that means 'before the nucleus'), because they have no nucleus. The cells of other living things are called **eukaryotes**, which means 'true nucleus'.

2 The cell cycle

The series of events that take place in a cell from when it divides to another division is called the **cell cycle**. For a cell to divide, it first needs to copy its essential parts. The most important part of the cell that needs to be copied is the **genetic material** in the **nucleus**. This provides the set of instructions for the new cell to look after, repair and in turn replicate itself. **Eukaryote** cells go through a series of changes as they copy the genetic material and then divide.

The cell cycle begins when a new cell forms, and it ends when the cell completes a division. How long this cycle takes varies from one cell type to another. Some cells divide more quickly. For example, cells that suffer a lot of wear and tear, such as those lining the small intestine in humans, might divide every eight hours or so. Roughly every five days, the small intestine lining is entirely replaced. Other cells divide more slowly. An average mammalian cell will divide about once every day.

The normal cell cycle of a plant or animal cell can be divided into two main phases: a growth phase, called **interphase**; and a cell division phase. The cell division phase has two parts, **mitosis** (the division of the cell nucleus) and **cytokinesis** (the division of the rest of the cell).

Interphase

Seen under a light microscope, a cell looks quiet during interphase. In fact there is a great deal going on. **Proteins** and other materials in the cell are being produced, and the **DNA** in the cell's nucleus is being copied. Also during interphase, the cell's **organelles** are copied, so that each of the two new cells will have all it needs to begin working right away.

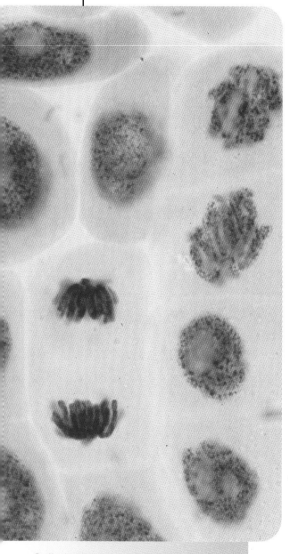

Cells in the growing root tip of a garlic plant (*Allium sativum*). The nucleus of the cell (lower centre) has divided, but the cell itself has not yet split. Magnification approx. x 1000.

Interphase is divided into three stages, called G1, S and G2:
- During the G1 phase the cell grows and prepares for DNA replication (copying).
- In the S phase a copy of the cell's DNA is made.
- In the G2 phase, the DNA has been replicated, and the cell prepares for division.

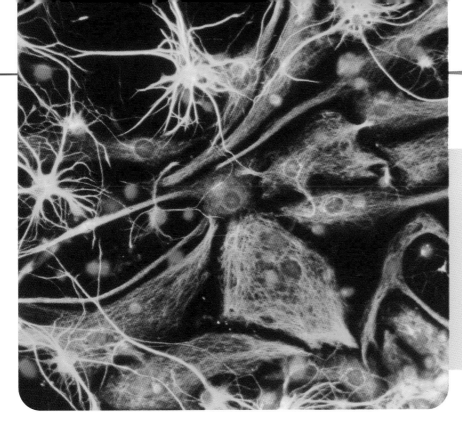

In most multicellular organisms, some types of cell become specialized and lose their capacity to divide. The star-shaped cells are specialized cells of this type. They supply nerve cells in the brain with nutrients. Magnification approx. x 350.

The time a cell spends in G1 or G2 can vary considerably. Cells such as nerve and muscle cells do not divide at all once they have become fully mature. They remain firmly in the G1 phase for months or even years, as they carry out their tasks in the body. In actively dividing cells, interphase is a period of great activity. During G1 the cell can double in weight. This is the time when new organelles are formed.

When the cell reaches a certain size, it enters the S phase. If for some reason the cell cannot reach the right size, for example because there are not enough nutrients for it to grow, then it does not divide. Instead it enters a resting phase until conditions improve. However, once it reaches the critical size it begins to copy its DNA and will carry on through the other stages of cell division.

During G2, the cell structures needed for mitosis and cytokinesis are assembled. In a non-dividing cell the DNA molecules are like very thin threads. During G2 these DNA molecules begin to coil. Eventually the molecules will form compact structures called **chromosomes**. The cell is now ready to begin mitosis.

Prokaryote cell division

The division of a **prokaryote** cell is a much simpler process than in eukaryotes. First, the genetic material, a single DNA molecule, is copied. Then both the original and the copy attach themselves to a different part of the cell membrane. When the cell begins to pull apart, the two DNA molecules are separated, forming two cells with the same genetic material. Because the process is much simpler, prokaryotes can divide more quickly than eukaryotes. Under ideal conditions, some bacteria can divide once every 20 minutes.

New nucleus

Every **eukaryote** cell has a **nucleus**, and every time a cell divides a new nucleus is formed, so that the **DNA** can be passed on to the next generation of cells. **Mitosis** is the division of the cell nucleus into two identical copies, each containing a complete set of **chromosomes** (a copy of the parent cell's DNA).

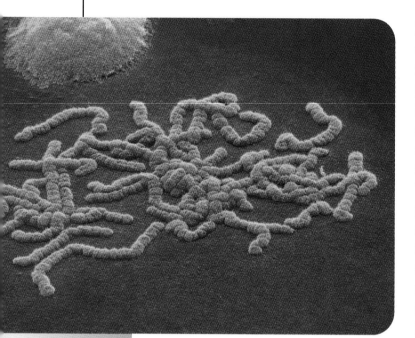

Mitosis begins after the cell has replicated its DNA and has finished constructing new cell parts. For convenience, mitosis can be divided into a number of stages, although these stages blend into one another in reality and there are no clear boundaries between them. The four stages of mitosis are **prophase**, **metaphase**, **anaphase**, and **telophase**. The whole process, from prophase to the separation of the two nuclei, telophase, usually takes between 30 minutes and three hours. Division of the cell itself (**cytokinesis**) follows.

The nucleus of this cell is about to divide. The DNA has coiled itself into a number of compact chromosomes. Magnification approx. x 1000.

Prophase

In **interphase**, the DNA in the nucleus is mixed up with **protein** in a tangled mass called **chromatin**. During prophase the DNA coils tightly upon itself, then coils even further around some of the proteins within the nucleus. This coiling causes the chromosomes to shorten and fatten, becoming thick, rod-like structures that are visible under the microscope. Each chromosome can be seen to consist of a pair of identical structures called **chromatids**, joined in the middle at a region called the **centromere**. This gives each pair a rough X-shape.

At the same time as the chromosomes are appearing in the nucleus, the cell is forming a structure called a **spindle**, made from thin tubes of protein called **microtubules**. Its job is to separate the two sets of DNA. In the spindle of an animal cell, the microtubules are attached to two pairs of stubby, rod-like structures called **centrioles**. One pair of centrioles moves to each end of the cell, each one with a fan of microtubules radiating out from it. At this point the membrane surrounding the nucleus starts to break down. This event marks the transition from prophase to the next stage of mitosis: metaphase.

Metaphase

During metaphase, the chromatids are moved to the centre, or equator, of the cell. The microtubules of the spindle become attached to the pairs of chromatids. Microtubules from both sides of the cell attach to the centromere of each chromatid pair. When the chromosomes are lined up at the cell's equator, anaphase begins.

Anaphase

At the start of anaphase, the microtubules of the spindle begin to pull apart the pairs of chromatids. One chromatid from each pair goes to one end of the cell, the second goes to the other end. These will be the chromosomes of the two new cells. Anaphase lasts only a few minutes. By the time it is completed, the spindle has disappeared and there are two identical sets of chromosones, one at each end of the cell.

Telophase

During telophase the new cell nuclei form. The chromosomes begin to unravel and become threadlike once again, disappearing from view as seen through a miscroscope. As this happens, new nuclear membranes begin to form around the chromosomes. Mitosis is now at an end.

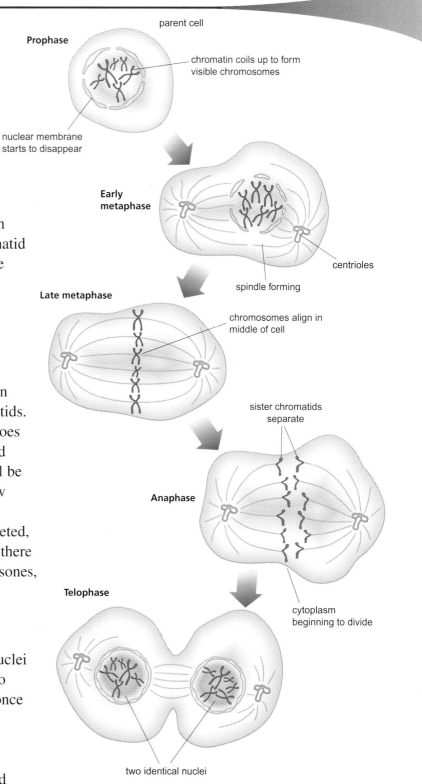

Prophase
parent cell
chromatin coils up to form visible chromosomes
nuclear membrane starts to disappear

Early metaphase
centrioles
spindle forming

Late metaphase
chromosomes align in middle of cell

Anaphase
sister chromatids separate
cytoplasm beginning to divide

Telophase
two identical nuclei

The four stages of mitosis: prophase, metaphase, anaphase and telophase.

Cell splitting

At some point between late **anaphase** and the end of **telophase**, the cell's **cytoplasm** (everything between the outer membrane and the **nucleus**) divides in two to produce two separate daughter cells. This is called **cytokinesis**. The way cytokinesis occurs differs between **organisms**.

Animal cell division

In animal cells, a process called cleavage takes place. All cells have a cytoskeleton (cell skeleton) – a network of thin protein fibres called microfilaments, which are attached to the membrane and give the cell shape. These microfilaments have the ability to be lengthened or shortened. As shown on page 9, at the start of cleavage, a band of microfilaments around the middle of the cell begins to pull in the cell membrane, producing a deepening furrow around the middle of the cell. This cleavage furrow continues to get deeper until it splits the cell in two.

The two new cells each have a complete cell membrane, and cytoplasm containing **mitochondria** and other **organelles** necessary for a successful life. Each cell has its own nucleus with a full set of **chromosomes (genetic material)**, containing all the instructions for growing and dividing.

Plant cell division

Plant cells differ from animal cells in that they have an outer cell wall in addition to the cell membrane. The cell wall is too rigid to be pinched together as in the cleavage of an animal cell. Instead, a new wall is formed down the middle of the cell.

This electron micrograph of kidney cells shows cleavage about to take place. Magnification approx. x 1000.

First, small sacs called **vesicles**, crammed with the materials necessary for building cell walls, are produced by the cell. These vesicles move to the equator of the cell, where they join with the remains of the **spindle**. The membranes of the vesicles begin to join up to form two new membranes across the middle of the cell. At the same time, the 'building materials' inside the vesicles form a disc-like cell plate between the two membranes. The cell plate grows until it joins up with the existing cell wall. The two new plant cell nuclei are now separated from each other by two membranes and cell walls.

Millions of divisions

It is the almost perfect repetition of the **cell cycle**, time after time, that makes us what we are. A single fertilized **egg** cell divides and divides until there are around 65 million million cells. As a result of this superbly accurate copying, every cell in the human body contains the same genetic material as the egg cell from which it grew.

The protistan *Plasmodium*, the microbe that causes malaria in humans. At one point in its life cycle, *Plasmodium* forms around 1000 offspring by multiple division. Magnification approx. x 3400.

Mutiple copies

Some of the simplest **eukaryote** organisms are single-celled creatures called **protistans**. A few protistans reproduce by multiple fission, in which the parent cell divides into a large number of tiny offspring. An example is the protistan *Plasmodium*, the parasite that causes malaria. When a type of malarial cell called a sporozoite infects a human liver cell, it divides by multiple fission to produce around one thousand offspring cells. These are then released into the blood, where they attack red blood cells, the cause of the fever associated with the disease.

3 Reproduction

The sequence of events that occur as a living **organism** grows, develops and eventually reproduces itself is called its life cycle. Growth and reproduction are always part of a complete life cycle.

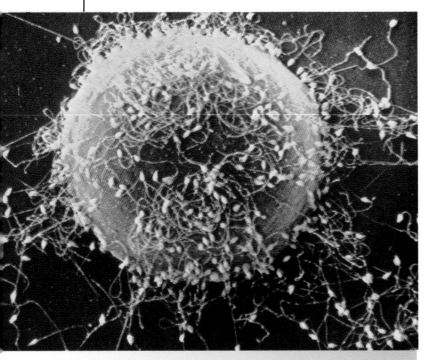

Sperm cells around a sea urchin egg cell. Only one of the sperm will penetrate the egg cell. Its nucleus will then fuse with that of the egg cell. Magnification approx. x 500.

Types of reproduction

Multicellular organisms can reproduce in one of two ways. Asexual reproduction is the equivalent of simple division in single-celled organisms. A single individual in some way produces one or more offspring. All the **genetic material** comes from the single parent, so parents and offspring are genetically identical – they are clones.

Most multicellular organisms reproduce at least part of the time by sexual reproduction. In sexual reproduction two individuals produce offspring that have genetic material from both parents. Sexual reproduction introduces new gene combinations in populations. Human beings and other living things come in two distinct varieties – male and female. In order for a new human to come into being a male sex cell, called a **sperm**, has to join with a female sex cell, called an **egg**. These sex cells (**gametes**) are produced in special reproductive tissues or organs. The offspring is a unique individual, and may have characteristics that are different from those of either parent.

Half the chromosomes

Every **species** has a certain number of **chromosomes** in its cells. These chromosomes exist in pairs. For example, human beings have 46 chromosomes in 23 pairs, while pea plants have 14 chromosomes in 7 pairs. The two chromosomes of a pair are not identical, but they are similar in size and shape. Most importantly, both chromosomes are responsible for producing the same characteristics in the cell. These pairs of chromosomes are called **homologous pairs**, meaning that each member of the pair is like the other. A cell with two of each type of chromosome is said to be **diploid**.

In sexual reproduction, the genetic material from two cells join together. If both these cells had the normal number of chromosomes, then the new cell would have twice as many chromosomes. In the next generation the chromosome number would double again, and so on.

Obviously the chromosome number cannot double in each new generation like this. To avoid this problem, when gametes form they have only half the usual number of chromosomes (one of each homologous pair). A cell with only one of each type of chromosome is said to be **haploid**.

Alternation of generations

In most plants, there are two distinct forms of the plant, occurring alternately. This is called alternation of generations. One of the forms of plant is diploid, and is called the sporophyte generation. This plant produces not gametes but tiny haploid cells called spores. Spores are resting cells that are good at surviving difficult conditions such as drought and extreme cold. Once conditions improve the spores grow and develop into the second form of the plant, called the gametophyte. The gametophyte produces male and female gametes, which join to produce a new generation of sporophytes.

Thus the life cycle of most plants alternates between producing spores and producing gametes. In simple plants such as mosses, the main form of the plant is the haploid gametophyte, with half the full chromosome number. The moss sporophyte is a tiny stalk and capsule that grows on top of the moss plant. In ferns, the parts that we see are the sporophyte plants, while the gametophytes are tiny. In flowering plants, which includes most plants and trees, the main plant is the sporophyte. The male gametes are formed inside the pollen tubes that grow from pollen grains, while the female gametes are egg cells contained in structures called carpels. The gametophyte is not a separate plant at all, but is hidden away in the flower.

An electron micrograph of pollen grains from a range of different flowering plants. Within each grain are haploid **nuclei**. The male gametes form from these nuclei after pollination. Magnification approx. x 300.

Meiosis

We have seen that the **gametes** formed for sexual reproduction are **haploid**: they have only half the usual number of **chromosomes**. When gametes are formed, a different kind of cell division takes place, one in which the number of chromosomes in the new cells is halved. This special division of the **nucleus** is called **meiosis**.

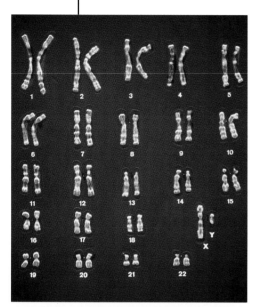

A photograph of the 23 pairs of human chromosomes. These are the chromosomes of a male: a female would have two X chromosomes rather than an X and a Y. Magnification approx. x 2800.

We have already learned that the chromosomes in a normal **eukaryote** cell are paired. The two chromosomes in the pair are not identical, but they do the same job. They are a **homologous pair**. Because of this pairing, it is possible for some cells with only half the chromosome number (a haploid cell) to function perfectly well. Indeed, as we have seen, plants such as mosses are haploid.

Gametes are formed from cells called germ cells, which are found in the reproductive organs. In meiosis a germ cell divides not once but twice. Before meiosis begins, the germ cell's **DNA** is duplicated just as it would be before **mitosis**. Exact copies of each chromosone are made, forming a joined pair of **chromatids**, and the cell then divides. The resulting cells then divide again immediately, without the DNA being copied beforehand. The result is four haploid cells.

First division

During **prophase** of this first division (meiosis I), the duplicated chromosomes pair with their homologues. This joining of homologous pairs is unique to meiosis: it does not happen during mitosis. As in mitosis the chromosomes coil and become visible under the microscope. The chromosones pair together. Each chromosone is formed from a pair of identical chromatids joined at the **centromere**. At this stage another process unique to meiosis can happen. The homologous pairs of chromosomes may swap segments of DNA with each other. This is known as crossing over. The crossing over process is an important way of producing a greater variety of characteristics in offspring. Following prophase is **metaphase** where the chromosomes move to the centre of the cell.

In the **anaphase** of meiosis I, the homologous pairs of chromosomes separate and move to either end of the cell to form the new nuclei in telophase. Division may now continue to produce two daughter cells. Each of these daughter cells has only half the normal number of chromosomes, but each of the chromosomes has two chromatids (so there are two identical copies of the DNA).

Second division

In meiosis II, the second division, the two daughter nuclei formed in meiosis I divide to form four new nuclei. Meiosis II is similar to ordinary mitosis, but there is often no interphase before this division begins (there is no need for one, because there are already two copies of the DNA).

In meiosis II the chromosome pairs move to the centre of the daughter cells. As they do so, a **spindle** forms at either end of the cell, and attaches to the chromosomes. The two chromatids in each chromosome are then pulled apart, and move to opposite ends of the cell.

By the time meiosis II has been completed there are four daughter nuclei in four separate cells. Each daughter cell has only half the number of chromosomes that the germ cells had, and it has only one copy of each of these chromosomes.

The stages of meiosis: meiosis I where the nucleus divides and homologous pairs of chromosomes are seperated, and meiosis II where the nucleus divides again, separating the two copies of each chromosome.

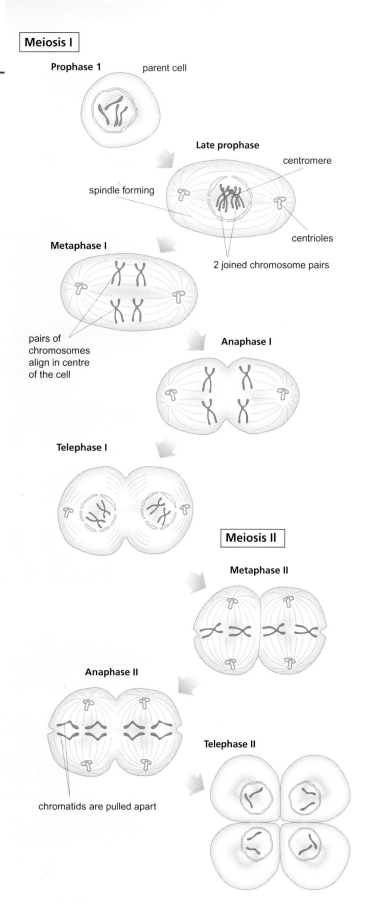

Meiosis I

Prophase 1   parent cell

Late prophase

centromere

spindle forming

centrioles

2 joined chromosome pairs

Metaphase I

pairs of chromosomes align in centre of the cell

Anaphase I

Telephase I

Meiosis II

Metaphase II

Anaphase II

Telephase II

chromatids are pulled apart

four nuclei with half the number of chromosomes

Fertilization

In all living **organisms** fertilization takes place when a male **gamete** and a female gamete join together. As we have seen both male and female gametes carry half the number of **chromosomes** (the **haploid** number) of the adult. When they fuse (join), a cell called a **zygote** is formed, which has the **diploid** number of chromosomes. This fusion of gametes is called fertilization.

Bringing gametes together

The way fertilization takes place varies from **species** to species. In many plants, for example, fertilization can take place between gametes from the same individual (self-fertilization) or between gametes from different individuals (cross-fertilization).

Bees are probably the most important pollinators of many flowering plant species. Honey bees are capable of carrying large amounts of pollen in 'pollen sacs' on their back legs.

Plant gametes

Primitive plants, such as mosses and ferns, require a moist environment so that the male gametes can swim across the plant to reach **egg** cells and fertilize them. In flowering plants the male gametes are carried in pollen grains to the **stigma**, the female part the flower. Pollen is most commonly transported to another plant of the same species either by insects or by the wind. This transfer of pollen from the male to the female parts of flowers is called **pollination**.

Pollen grains that succeed in reaching a receptive stigma begin to grow a microscopic pollen tube that penetrates down to the ovule. A **nucleus** from the pollen grain travels down the pollen tube to the ovule and fuses with the nucleus of the egg cell.

A female common frog (*Rana temporaria*) laying her eggs, or spawn. In most frog species the eggs are fertilized externally, after they have been laid.

Animal gametes

In animals, fertilization is usually between different individuals. Fish and amphibians release their eggs and **sperm** into water and the sperm swim to the eggs to fertilize them. This is called external fertilization. In land-living animals – reptiles, birds and mammals – the male inserts his sperm through an opening in the female's body. This is internal fertilization. The advantage of internal fertilization is that mating does not need to take place in a wet environment for the sperm to swim to the egg cells.

In male animals the gametes (sperm) are very small, are produced in huge numbers, and develop flexible 'tails' that enable them to swim towards an egg. Different species produce different types of sperm cell. The sperm of an opossum appear to have two tails, for example, because pairs of sperm swim together for greater efficiency.

In contrast to sperm, eggs are very large, are produced in small numbers and are unable to move. The egg has all of the cell material and structure it needs to get the new individual off to a good start if fertilization takes place.

The male sperm swims towards the female egg and penetrates its outer layer. It then releases its nucleus into the egg cell, and this joins the egg's nucleus in a single diploid nucleus. This is the first cell of a new organism and is called the zygote.

4 Development

When a male and a female **gamete** fuse to form a fertilized **egg** (the **zygote**), it is the beginning of a potential new **organism**. But this is only the first step in a long and complex journey from that single-celled zygote to a multicellular organism. The process is called development, and the instructions telling the cell how to develop are held in the organism's **genetic material** – its **DNA**. Development involves shaping the organism, and the specialization of groups of cells to carry out particular tasks.

Animal development

Development in animals begins when the zygote (a new **diploid** cell) starts to divide repeatedly. Once it begins dividing, the zygote is called an **embryo**. At this stage the contents of the original egg cell are simply being divided up among a growing number of smaller cells, each with its own **nucleus**. There is no cell growth and the embryo stays at about the same size.

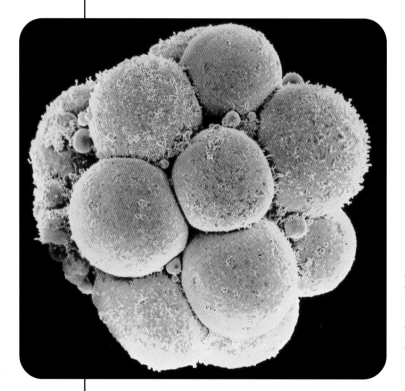

At first the embryo consists of a ball of identical cells. This stage is called cleavage. Eventually a fluid-filled, hollow ball of cells called a **blastula** forms. The frantic pace of cell division slackens off and the embryo enters the next stage of development.

A human embryo four days after fertilization. The embryo is still a ball of very similar cells. Soon these cells will begin to develop into different cell types. Magnification approx. x 900.

Germ layers

In the next stage of development, the blastula goes through a major reorganization. The cells of the blastula rearrange themselves to form a **gastrula** of two or three layers of cells, one inside the other. The cells that develop from these layers – the germ layers – will form all the tissues and organs in the adult organism.

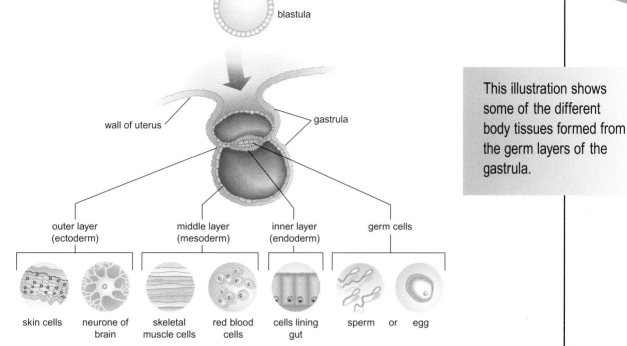

The illustration shows the stages from zygote (fertilized egg) to blastula to gastrula, with labels for wall of uterus and gastrula, and the germ layers:

- outer layer (ectoderm): skin cells, neurone of brain
- middle layer (mesoderm): skeletal muscle cells, red blood cells
- inner layer (endoderm): cells lining gut
- germ cells: sperm or egg

This illustration shows some of the different body tissues formed from the germ layers of the gastrula.

The outermost of the three layers of cells is called the **ectoderm**. From this layer, all the cells of the skin and of the nervous system develop. Within the ectoderm is the **mesoderm**, or middle layer. This layer produces the muscles and most of the skeleton, as well as the heart and blood vessels, the kidneys, and the reproductive organs. The innermost layer of cells, the **endoderm**, forms the lining of the gut and the organs associated with it.

Once the germ layers have formed, the cells in each layer quickly become distinct from one another, as different groups of cells, or tissues, begin to specialize for particular tasks. This is the beginning of organ formation. Once the organs have formed, they grow and begin to function, taking on the jobs they will perform throughout the life of the organism.

Guiding genes

One of the important areas studied by biologists is how a fertilized egg, one single cell, can develop into an organism with many millions of cells, of many different types. The development process is directed and regulated by the DNA in the cell nucleus. Sections of the DNA, called genes, interact with one another to control development. Only a few genes concerned with the essential processes in the cell are active all the time. Other genes are only turned on at particular stages in development.

Growth and development

If an organism is going to get bigger, its cells have to divide and make more cells. This is growth. Development, on the other hand, is the appearance of different, specialized body parts.

19

Plant development

The description of plant development below relates mainly to **angiosperms** – the flowering plants. The vast majority of the plants around us are angiosperms.

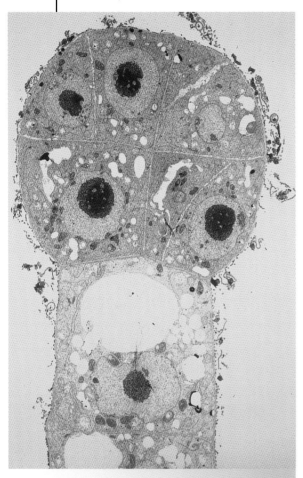

This dissected embryo of the turnip (*Brassica campestris*) shows the cell (bottom) that will develop into the first root, and the cells (top) that will develop into the shoot and leaves. Magnification approx. x 500.

As soon as fertilization has taken place, the new **zygote** begins to divide. As in animals, it forms a group of cells, which then specialize to form an **embryo**. The embryo develops the beginnings of a young shoot, which will form the plant's stem and leaves, and a young root. The embryo also forms one or two seed leaves, or **cotyledons**.

The new embryo remains at the centre of the ovule, the part of the flower where the **egg** cell originally formed. It is surrounded by a specialized tissue called the **endosperm**, which supplies the growing embryo with food. The ovule itself develops into a seed and forms a tough protective seed coat around the embryo and endosperm. While this happens, parts of the flower surrounding the seed develop into a fruit. As the fruit ripens, the seeds lose water and become quite dry.

Spreading the seeds

Once seeds are fully formed, development of the embryo inside stops, and the seed may become dormant. Before the plant can develop further, the seeds must be scattered from the parent plant.

Different plants spread their seeds in different ways. Some fruits are juicy, to attract animals to eat them. If they are eaten, the seeds survive the journey through the animal's gut and grow in its droppings. Some plants have very light seeds that are carried by the wind. In some cases the seeds are attached to parachutes or wings that help them travel on the wind. Some fruits are pods, which burst open and shoot the seeds out. There are also seeds that are carried by water, and seeds that hook on to passing animals.

Once a seed comes to rest, it will remain dormant until conditions are right to trigger the next stage in the plant's life: **germination**.

Germination

Germination is the restarting of growth of a plant embryo inside its seed after a period of **dormancy**. The most essential ingredient for germination to begin is water. The seed takes up water through a tiny hole in the seed coat, causing it to swell up. As a result of this the seed coat bursts open. Once the seed has split, the young shoot begins to grow up towards the light and the young root grows into the ground. The shoot is bent into a hook shape, to protect its delicate growing tip as it pushes up through the soil.

The food source for the early growth of the embryo comes from the seed leaves (the cotyledons). In plants such as broad beans, the cotyledons remain underground, inside the seed coat. Food stored in the cotyledons is converted into sugars, which can be transported to the shoot and root tips where growth is taking place. In other plants the cotyledons are lifted up out of the soil and start to make food by **photosynthesis**. Once the new seedling grows its first green leaves, the cotyledons wither away.

Germination of beech tree seedlings. The leaf-like structures on the seedling on the right are the cotyledons.

Growing places

Plant growth is limited to specific growing zones called **meristems**. These are areas of unspecialized cells found at the tips of roots and shoots, and as a layer of cells in the stems and roots. Plants are shaped through different patterns of growth at the meristems.

Arrested development

The record for seed dormancy is held by the Arctic lupin. Seeds found in the frozen earth in the Yukon, Canada in 1954 were estimated to be between 10,000 and 15,000 years old. The seeds germinated successfully in 1966.

Is there an internal aging clock in every **organism**? Fruit flies live for about a month, dogs for about 15 years, humans perhaps over a hundred years if they are very lucky. The consistency of lifespan in each **species** suggests that genes are involved.

Throughout an organism's life it is aging. Changes in the structure of the organism and in its workings, especially after sexual maturity has been reached, affect every organ of the body. Eventually these changes cause it to deteriorate physically and die. There are many theories about how and why aging occurs, but none yet accounts for all the known facts about the process.

Aging cells

In mammals, there are many changes as the body gets older. One of the main effects is a steady loss of muscle. However, as the muscles waste away, the amounts of fluid and stored fat often increase, so that overall the body becomes heavier. The body's **metabolic rate** (the speed of the chemical processes in the body's cells) gets slower, which reduces the amount of heat it produces. To make up for this, the body reduces the blood supply to the skin, to cut down on heat loss. The bones of the skeleton become weaker and are more easily broken. The skin becomes less elastic, which causes wrinkles to form.

As a person ages, changes in the skin make it less elastic, which leads to wrinkles.

Some types of body tissue are continually renewed by cell division. The cells of the liver and pancreas, for example, are constantly being replaced by new cells, so the effects of aging are not so severe in these organs. Highly specialized tissues such as nervous tissue, however, do not divide once they are mature. These tissues show definite signs of deterioration as aging progresses. For instance, the number of nerve cells in the nervous system and in the brain falls steadily. Overall, cell division gets slower as the organism ages. Wounds take longer to heal, and infection is harder to fight off because the **immune system** produces fewer cells.

Plants also age. Some kinds of plant seem to be genetically 'programmed' to die after one or two years or even a few months. Although other kinds of plant can grow for years, their ability to do so declines with age.

Genes and aging

Studies of human cells cultured in the laboratory have revealed that a normal cell can usually divide only about 50 times before it dies. Most cells in the human body divide no more than eighty or ninety times. This limit is thought to be controlled by structures called **telomeres** on the tips of **chromosomes**. Each time a cell divides, part of the telomere on each chromosome is lost. When the telomere is lost completely, the cell dies.

Some people have suggested that the evolutionary value of aging lies in the act of giving the next generation a better chance to evolve and adapt to changes in the environment. This suggests that natural selection has favoured genes that cause aging. Reproductive success in any organism means producing offspring – setting the next generation on its way.

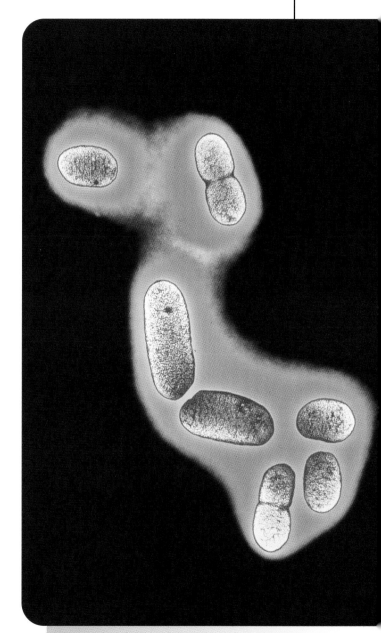

Immortal cells

Some cells never lose their ability to divide. Examples are sex cells and cancerous cells. In these 'immortal' cells, **enzymes** rebuild the telomeres after each cell division.

Bacteria (*Escherichia coli*) in the process of dividing. Bacterial chromosomes are circular, and so they do not have telomeres on the ends of the chromosome. As a result they do not age like most **eukaryote** cells. Magnification approx. x 10,000.

5 Genes and DNA

Each kind or **species** of **organism** has a set of characteristics that are passed on from generation to generation. A blackbird, for instance, will develop feathers and will be able to fly, while a beech tree will develop a trunk, bark and leaves. However, individuals within each species are not exactly the same. There is variation in their characteristics. A human may be tall or short, have blue eyes or brown eyes, be male or female.

All these different inherited characteristics are carried by **DNA**, the cell's **genetic material**. DNA is the instruction manual that says this **egg** will develop into a human and not a halibut, this seed will grow into an oak and not an orchid. It also influences whether a person will be tall or short, fair-haired or dark.

There are many questions that we can ask about the genetic material and how it works. What exactly are the 'instructions' that are passed on from generation to generation? How do they produce their effects? Also, how are these instructions copied faithfully every time a cell divides?

By looking more closely at DNA, we can answer these questions.

The characteristics of these kittens are determined by the genes of their mother and their father.

In the genes

The basic unit of inheritance, by which characteristics are passed from one generation to the next, is called the gene. When the word gene was first used at the beginning of the twentieth century, it referred to something passed from one generation to the next that determined a particular characteristic, such as the colour of a person's eyes or the height of a pea plant, but at that time, no one was sure what a gene was.

A **chromosome** is a very long strand of DNA, wrapped around **proteins**. We now know that a gene is one length of DNA. Genes are arranged along the chromosome like beads on a necklace. Each chromosome is likely to have thousands of genes. Within a species, the gene for a particular characteristic will always be found at the same place on the same chromosome.

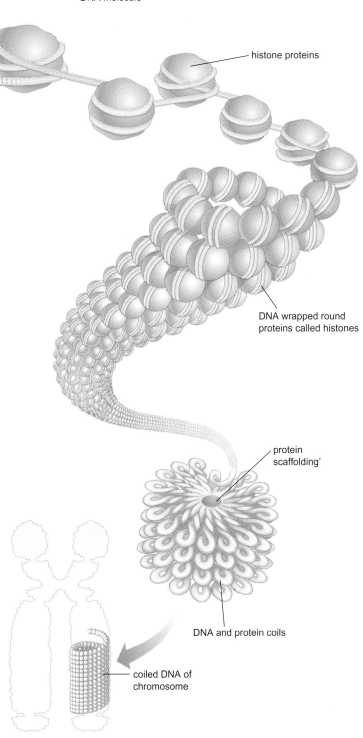

DNA molecule

coiled DNA

histone proteins

DNA wrapped round proteins called histones

protein scaffolding'

DNA and protein coils

coiled DNA of chromosome

A simplified model of a chromosome's structure, showing how DNA coils around itself and then around a 'skeleton' of protein molecules.

Cells have two copies of each gene, one from each parent. These two versions of the same gene are called **alleles**. The particular combination of alleles in an individual is what makes them unique. With thousands of genes on each chromosome, the number of different combinations of alleles from one parent with alleles from another is almost countless.

To learn more about genes and how they work, we need first to understand more about the structure of the materials that chromosomes and genes are made of – DNA.

DNA – the molecule of life

In the early 1950s James Watson and Francis Crick, two scientists working at Cambridge University in the UK, worked out the structure of the **DNA** molecule for the first time. Crick was so thrilled by the discovery that he ran into a local pub announcing 'We've discovered the secret of life!'

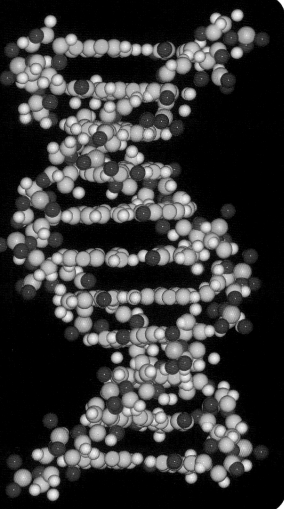

A computer model of part of a DNA molecule, showing its 'twisted ladder' shape. The coloured spheres represent individual atoms.

More than any other substance, DNA defines something as being alive. Nothing lives without DNA. Every single living cell, without exception, possesses DNA at some point in its life. Built into the make-up of this molecule are the instructions for constructing and maintaining cells.

The structure of DNA

Watson and Crick's model of DNA showed that a DNA molecule is rather like a twisted ladder – a double helix. It is made up of two immensely long chains, each composed of many smaller subunits called **nucleotides**. The chains are connected at regular intervals to make the 'rungs' of the ladder.

These rungs are made up of pairs of chemicals called **bases**. There are four such bases in DNA, called adenine (A), cytosine (C), guanine (G) and thymine (T). One of the first clues to working out the structure of DNA was that the amount of cytosine present in the molecule always equalled the amount of guanine, and thymine always equalled adenine.

DNA duplication

We saw earlier that a cell duplicates its **chromosomes**, and therefore its DNA, before it divides. How is this done? We have seen that DNA is not a single molecule, but a double one. The double chain means that DNA has one characteristic that is essential to the continuation of life: it can reproduce itself.

Rather than thinking of DNA as a twisted ladder, think of it as a twisted zipper, with the pairs of bases as the interlocking teeth of the zipper. As we have seen, the bases on the DNA chains always pair up in the same way. Adenine is always paired with thymine, and guanine always pairs with cytosine.

When DNA is duplicated, the first thing that happens is that the two strands of DNA separate from one other – the zipper is unfastened. **Enzymes** bring more nucleotides to the two unzipped DNA strands. New nucleotides link up with the exposed bases on the unzipped strands, A to T and G to C. The new nucleotides then join together to form new strands attached to the original strands. The result is two double strands of DNA, rather than one.

The ability of DNA to replicate (copy itself) in this way is essential to cell division and to reproduction. The genetic material has continued to be copied, from generation to generation, in an unbroken thread running back through time. This DNA connection goes back beyond dinosaurs and trilobites to the very first cells.

Rosalind Franklin

In 1951, the scientist Rosalind Franklin used a technique called X-ray diffraction to obtain images of the DNA molecule. She directed a beam of X-rays at a crystal of DNA and captured the image on film. She then used this to work out the positions of the atoms in the molecule. Her results strongly suggested that the molecule had a helical structure. In 1953 Watson and Crick used her pictures to work out that DNA spirals into a double helix. Watson and Crick were awarded the Nobel Prize in 1962 for this work. Sadly, Franklin did not share in this as she died of cancer, aged 37, and the Nobel prize is never given posthumously.

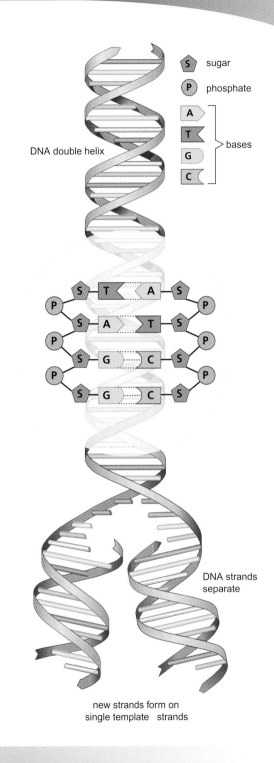

DNA double helix

S sugar

P phosphate

A
T
G
C
} bases

DNA strands separate

new strands form on single template strands

The structure of the DNA molecule, and how it replicates (copies itself).

The genetic code

The structure of **DNA** contains a four-letter code of instructions (see box at bottom of page 29) – but what for? The instructions are recipes for making **proteins**, the substances that control the day to day running of cells.

The cell's managers

Proteins are key substances in every **organism**. Many of the structures in the body are made of proteins. For example muscle fibres are proteins, and proteins are part of every cell membrane.

Other kinds of protein are even more essential to life. Proteins called **enzymes** are natural catalysts – substances that control the rate of chemical reactions in the cell. There are thousands of such reactions happening in a cell at any one time, as molecules are broken down to release energy, or built up to form the molecules that make up the living cell.

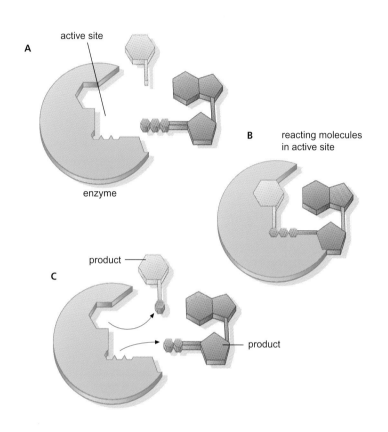

A active site

enzyme

B reacting molecules in active site

C product

product

Enzymes increase the speed of many cell reactions. Others would not happen at all without the enzyme. The shape of the enzyme molecule has a space called the 'active site' (A), where the substances involved in the reaction fit exactly, like a key in a lock (B). After the reaction the new products are released from the active site (C).

Without enzymes there would be no life. A healthy cell needs to have all the right enzymes to perform essential tasks, such as getting energy, moving, growing and repairing itself. Each reaction that takes place in the cell has its own specific enzyme. Because the enzymes control the reactions, they effectively control the cell.

Proteins and DNA

Proteins are complex substances. Protein molecules are long chains, built from smaller molecules called **amino acids**. There are about 20 different amino acids used in making proteins. Each type of protein has a specific recipe: a fixed order of amino acids along its length. This recipe has to be followed exactly, every time, or the protein will not be able to carry out its job.

We can now see why DNA is so essential to the cell. It is the instruction manual that holds the combinations for all the proteins that an organism makes. The protein-making instructions are held in the sequence of the **bases** along the length of the DNA molecule.

The genetic alphabet

As we have seen, there are four bases in DNA, adenine (A), cytosine (C), guanine (G) and thymine (T). The four bases, A, T, C, and G, are the alphabet of the genetic code. There are about 20 different amino acids, but only four bases in DNA, so how can four bases be a code for 20 amino acids? The answer is that each amino acid is coded for by not one base, but a group of three. These groups are called base triplets. The triplet AGC, for example, is the code for an amino acid called serine.

Now we can begin to see how DNA is used in the cell. The sequence of bases along the DNA molecule is a series of triplets. Each triplet codes for an amino acid, and a sequence of triplets codes for a protein. But there is still a very important question to answer. How does this code of bases on the DNA get translated into actual protein molecules?

The Human Genome Project is a huge research programme which is working to map out the complete genetic code for the human genome (all 23 chromosomes). This scientist is looking at a tiny portion of the sequence.

Duplicate codes

If you work out how many ways there are to arrange four letters in groups of three, you will find that there are 64. This means that there are 64 possible ways to arrange the four DNA bases in groups of three: the genetic code contains 64 'triplet codes'. Since there are only 20 amino acids, this means that there are 44 'spare' triplet codes. Scientists have discovered that some amino acids are coded for by more than one triplet: some have two codes, some have four, and some have six. There are also three triplets that act as 'stop' signs, marking the end of a protein sequence. There is also a triplet that acts as a 'start' signal, indicating the beginning of a stretch of code.

DNA makes RNA makes proteins

We have seen that the basic unit of inheritance, the gene, is actually a length of **DNA**. Now that we have learned more about the structure of DNA and the genetic code, it is possible to see that each gene is a section of DNA that codes for a **protein** (or for part of a protein). There are thousands of genes on each of the cell's **chromosomes**. So the cell's DNA is not just an instruction manual full of protein combinations, but a vast library, containing codes for thousands of different protein sequences.

A cell's DNA never leaves the **nucleus**. One reason is that the DNA molecules are too big. Each chromosome in the nucleus is a single DNA molecule. Although chromosomes vary a great deal in size, they are all very large, containing millions of **base** pairs. However, experiments show that proteins are made outside the nucleus, in the cell **cytoplasm**. So there must be a mechanism for copying parts of the DNA code and carrying them out of the nucleus.

Reading the code

During **interphase**, when a cell is not dividing, there is a great deal of activity in the cell nucleus. In many locations, sections of the double-stranded DNA are unravelled to allow a process called **transcription** to take place. What is happening is that sections of the DNA – individual genes – are being copied. As when DNA copies itself, free **nucleotides** line up and join to the DNA strand, pairing with the bases on the DNA. In this case, however, a fairly short stretch of the DNA is copied, and the molecule formed is not DNA but a close relation called **RNA** (ribonucleic acid).

How proteins are made. As messenger RNA passes through a ribosome, transfer RNA molecules bring amino acids to join on to the growing protein chain.

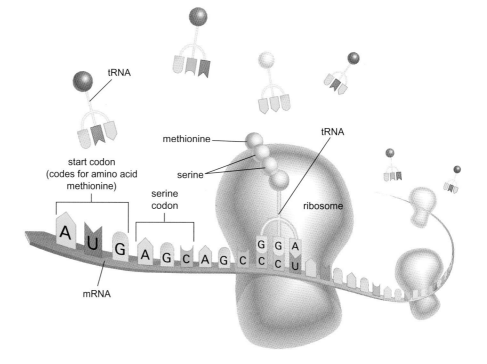

tRNA

start codon
(codes for amino acid
methionine)

serine
codon

methionine

serine

tRNA

ribosome

mRNA

There are several different types of RNA. One, called **messenger RNA** or mRNA, carries a copy of the code for a particular protein chain. Messenger RNA is similar to DNA, but has only one strand, rather than two. Also, one of the bases in RNA is different from the bases in DNA: instead of thymine (T), RNA has uracil (U).

Messenger RNA is so called because it carries the code for a protein chain out of the nucleus and into the cytoplasm. In the cytoplasm, the code on the mRNA molecule is 'read' by tiny bead-like structures called **ribosomes**.

Assembling a protein

Making a protein in the cytoplasm is rather like a production line in a factory. Ribosomes attach themselves to a mRNA molecule at the 'start' code for a particular protein. **Amino acids** are brought to the ribosomes by another type of RNA called **transfer RNA** (tRNA). Transfer RNA molecules are like fork-lift trucks, bringing the parts of the protein chain to the assembly line, ready to be fitted together. Each tRNA molecule has one end, that carries one particular kind of amino acid, and another end joins up with the code for that amino acid on the mRNA molecule.

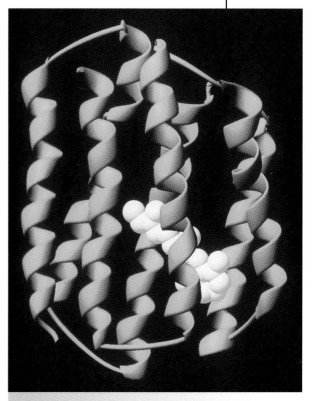

Once a protein chain has been assembled, it twists and folds up to form a complex three-dimensional shape. This computer graphic shows the protein rhodopsin, one of the light-sensitive pigments in the eye. The blue is the protein chain, while the yellow is a light-sensitive compound called retinal.

Let's say that a triplet on the mRNA is AGC, which we know codes for the amino acid serine. A tRNA molecule carrying the amino acid serine attaches itself to the mRNA. The ribosome then moves along the mRNA to the next triplet. If this is also AGC, a second tRNA carrying a serine molecule will attach to the mRNA. The two amino acids are now joined together by an **enzyme**. The first tRNA then drops off, leaving the two serine molecules attached to the second tRNA. The ribosome now moves along to the third triplet on the mRNA, and the process continues. As the mRNA code is read, triplet by triplet, a long chain of amino acids is built up.

It is possible for more than one ribosome to be moving along the mRNA at the same time as the cell busily assembles proteins. This is the genetic code at work.

Mutations

DNA copying is an incredibly accurate process. The DNA in a human cell contains about 3 billion **bases**. Cells divide time after time, and each time these 3 billion bases are copied correctly. However, very occasionally a mistake is made. The altered genes that result are called **mutations**. Usually cells that carry mutated genes die because they cannot function properly.

If the mistake happens as a normal cell is dividing, it will not have a great effect. The cell that has the mutation will probably die, and even if it does not it is only one abnormal cell among millions of healthy ones.

The situation is different if the mutation happens during the formation of a **gamete** (a sex cell). If the mutated cell survives, it may go on to join with another gamete and form a new individual. The mutation may then be passed on to the next generation. A mutation that is passed on can cause serious problems; in humans, for example, a blood problem called sickle cell disease is the result of a mutation. People who have this disease have large numbers of abnormal, sickle-shaped blood cells which are very poor at carrying oxygen. Other mutations may not have a very significant effect: many people are colour-blind, for instance, a condition that was originally produced by a mutation.

On rare occasions, a mutation can be beneficial and give the **organism** an advantage over its competitors. An example is when a mutation in a disease-causing bacterium gives it resistance to a drug. (This mutation is beneficial to the bacterium but not, of course, to humans with the disease).

Chromosome mutations and gene mutations

A mutation can affect an individual gene or an entire **chromosome**. A chromosome mutation involves a major alteration to one or more chromosomes, perhaps even a change to the number of chromosomes present. People who have the mental and physical disorder called Down's syndrome have been born with an extra copy of one chromosome.

The characteristic sickle or crescent shape of red blood cells in a person with sickle cell anaemia. Magnification approx. x 5200.

A gene mutation is a result of chemical changes in the cell's DNA. When a mutation occurs the order of DNA bases making up a gene is altered. This change can be very small but it can have profound consequences. Sickle cell disease, for example, affects haemoglobin, the **protein** in red blood cells that carries oxygen. It is caused by the alteration of a single base pair in the genes that code for haemoglobin. This alteration results in one **amino acid** in the haemoglobin being altered, causing the haemoglobin molecule to be distorted.

A rare event

Most mutations are the result of chance error when DNA is copied. Such an error is a rare event – a mutation will happen perhaps once in every 100,000 times that a DNA molecule is copied. However, in fast reproducing organisms such as bacteria, some of which can divide every 20 minutes in ideal conditions, there are more chances for mutations to occur. If a few drug-resistant individuals appear in a population of bacteria, there will soon be a whole drug-resistant strain, as the susceptible individuals are killed and the resistant individuals are left to breed alone.

The likelihood of mutations occurring is increased by exposure to such things as ultraviolet light, X-rays and some chemicals. Agents that cause mutations are called mutagens. There is no way to predict the type of mutation that will be caused by a mutagen.

Mutations and evolution

Most mutations that cause a visible change to the organism are harmful. However, on rare occasions, a mutation will improve an organism's chances of survival. If this individual then reproduces, it might pass this benefit on to its offspring. The offspring will also have improved chances of survival, and gradually, over many generations, the mutation will be found in most members of the **species**. This kind of mutation is one of the processes involved in evolution, the way in which organisms gradually adapt and change over time.

a) Photograph of a mutant fruit fly (*Drosophila melanogaster*) with four wings instead of the usual two. Fruit flies are widely used in genetic research because the DNA in their salivary glands is replicated many times to form giant chromosomes.

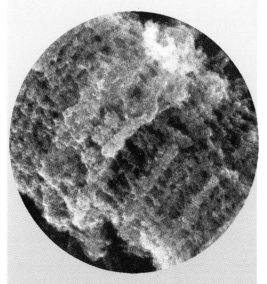

b) Patterns of bands in these giant fruit fly chromosones can be used to identify different parts of the chromosome.

33

6 Patterns of inheritance

Do you have a nose just like your mother's? Are you dark-haired, like your father? We have seen that most cells contain **homologous pairs** of **chromosomes**. One of these chromosomes comes from the female **gamete** (from the **organism's** mother), the other from the male gamete (from the father). Each chromosome carries many different genes. So an organism has two versions of every gene, one from its mother and one from its father.

These two versions of the gene may not be the same. So what happens when you inherit two different versions of the gene for a certain characteristic?

Dominant and recessive genes

Different versions of the same gene are known as **alleles**. Suppose you inherit a gene for brown eyes (we could call it B) from both your mother and your father. In this case you will have two identical alleles, BB, and your eyes will be brown. Similarly, if you inherit a blue-eyes gene (b) from each parent, you will have two b alleles, and your eyes will be blue. But what happens if you receive an allele for blue eyes from one parent and an allele for brown eyes from the other? The answer is that you will have brown eyes. The reason for this is that the allele for brown eyes takes over: it is said to be dominant over the allele for blue eyes, which is said to be recessive. The effect of the B allele is to mask the b allele so that no effect from it is seen.

Mendel's experiments

The founder of **genetics** was an Austrian monk called Gregor Mendel (1822–1884). He spent many years studying the inheritance of characteristics in pea plants. His experiments provided clear illustrations of some basic rules of genetics.

An illustration showing Mendel's experiment crossing white-flowered with purple-flowered pea plants.

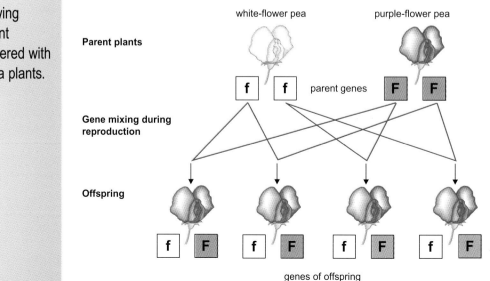

One of Mendel's experiments involved breeding together pea plants with specific characteristics and looking at the offspring. In one experiment, Mendel bred together pea plants that always produced white flowers and pea plants that always produced purple flowers. He found that all the offspring of this cross-breeding had purple flowers.

Today, we can explain the results of Mendel's experiment by saying that the purple-flowered plants had two identical alleles for purple flowers (FF). Similarly, the white-flowered peas had identical alleles for white flowers (ff). When the two plants were bred together, all the offspring had two different alleles, Ff, but the purple-flower allele was dominant, so all the plants had purple flowers.

Mendel's Laws

From his observations of pea plants, Mendel set out two laws of heredity: the first said that hereditary characteristics are determined by separate units (what we now call genes). These units occur in pairs. The genes in a pair segregate (separate) during the formation of gametes, and each gamete receives only one member of the pair.

Mendel's second law predicts that different characteristics will be inherited independently of each other. For instance, pea plants can be short or tall, and have white flowers or purple flowers. These characteristics can be inherited in any combination, to produce tall, white-flowered plants for instance, or short, purple-flowered plants.

Today, we know that things are not as simple as Mendel's laws suggest. We know that many characteristics are controlled not by a single gene but by a number of genes acting together. We also know that Mendel's second law is only true in some situations. It is only genes that are on different chromosomes that are truly independent of each other. Two genes that are close to each other on the same chromosome will often be inherited together.

A 'black panther' is a leopard with a black coat, not a separate species. A **mutation** is thought to have originally produced the black colour. In forests, this colouring is an advantage, and black panthers are common.

Phenotype and genotype

Genes alone don't dictate appearance. Genes interact with one another and with the environment to produce the outward form of the **organism** that we see. The actual appearance of an organism is called its **phenotype**. Underlying this phenotype are the genes that influenced this appearance – the **genotype**.

Homozygous and heterozygous

When the two **alleles** for a particular characteristic are the same, we say that the organism is **homozygous** for that characteristic. If both alleles are dominant then the organism is described as homozygous dominant, if both are recessive then it is homozygous recessive.

If, on the other hand, the two alleles are different, the organism is **heterozygous**. If one member of the heterozygous pair is dominant over the other then the organism's phenotype will be just the same as if it were homozygous dominant.

Breeding true

In his pea plant experiments, Mendel was careful to make sure that his plants always showed the characteristic he was interested in, for generation after generation. When all the offspring have the same characteristics as the parents this is called breeding true for that characteristic. Homozygous plants and animals will always breed true.

These prize-winning dairy cows have been bred for maximum milk production.

Farmers are particularly interested in making sure that the animals that they breed are as good, if not better, than previous generations. If, for example, cows give a good milk yield, the farmer will want their calves to give a good yield too. If a high-yield cow MM mates with a bull that also carries these homozygous alleles then all the offspring will be MM as well. However heterozygous animals, Mm, do not breed true.

Of course, this is a very much simplified example, as there are likely to be several genes controlling milk production, plus environmental factors such as how the cattle are fed. In a simple case, where a characteristic is controlled by one pair of alleles, what does happen when we breed two heterozygous organisms together? We can find out by looking at another of Mendel's experiments.

Second-generation peas

After his experiment breeding together purple-flowered and white-flowered peas, Mendel did a follow-up. He bred together the purple-flowered offspring from the first experiment.

In modern terms, these plants were all heterozygous for flower colour, Ff. If we think about what happens when the plants breed together, we can get some idea of what the results of this experiment might be.

Each of the pea plants will produce **gametes** with one flower colour gene. Half of these gametes will carry the F allele, and half the f allele. When the plants breed, there are four possible results. The offspring could receive a dominant allele from both parents (FF), or a recessive allele from both parents (ff). They could get an F allele from the mother and an f allele from the father (Ff), or vice versa (fF).

Of these four possible genotypes, three will produce purple flowers, but one (ff, the homozygous recessive) will produce white flowers. So for every four offspring, three should have purple flowers and one white flowers. Mendel's real-life experiment did not produce quite such clear-cut results, but over many repeated experiments on thousands of pea plants, he did indeed get close to this ratio of three to one.

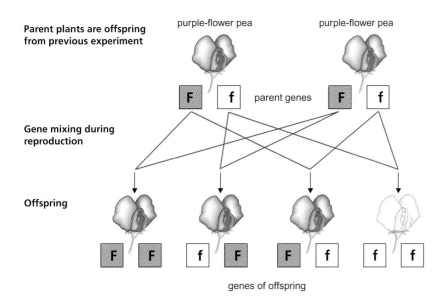

Parent plants are offspring from previous experiment

purple-flower pea   purple-flower pea

F   f   parent genes   F   f

Gene mixing during reproduction

Offspring

F   F   f   F   F   f   f   f

genes of offspring

An illustration showing a second experiment in which Mendel crossed the heterozygous purple-flowered pea plants from the first experiment.

37

Is it a boy or a girl?

What is it that determines your sex? Whether you are a male or a female depends on a particular pair of **chromosomes** called the sex chromosomes. Females have a pair of identical sex chromosomes, called the X chromosomes. In males the sex chromosomes are different. One is an X chromosome, as in females, but the other is a shorter Y chromosome. This means that females are XX and males are XY. In the case of the sex chromosomes, for once the **genotype** has a strong bearing on the **phenotype**!

When males and females form **gametes** (**sperm** and **eggs**), the sex chromosomes are separated during **meiosis**, along with all the other chromosomes. In the female, every egg produced has one X chromosome. However, in the male 50 per cent of the sperm will have an X chromosome and 50 per cent a Y chromosome. If a Y sperm fertilizes an X egg the result is XY, a boy. If an X sperm fertilizes the egg, the result is XX, a girl. Because there is a 50 per cent chance of either event happening the ratio of males to females in the population is more or less half and half.

Sex linkage

The sex chromosomes are different from all other chromosome pairs because the Y chromosome is shorter than the X chromosome. In fact, the Y chromosome has very few genes at all. So in a male, most genes on the X chromosome do not have a matching partner: there is only one **allele** for each gene.

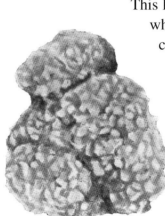

This lack of a second allele explains why there are some inherited conditions that are much more common in males than in females. In humans, colour blindness and the more serious condition of haemophilia (a disorder of the blood-clotting mechanism) are examples of this sex linkage.

An electron micrograph showing an X and a Y chromosome. Men have this combination, while women have two X chromosomes. Magnification approx. x 30,000

Queen Victoria was a carrier of the blood disease haemophilia. Her son Prince Leopold (second from the right) suffered from haemophilia, and several of her daughters and grand daughters carried the disease.

The allele that causes red-green colour blindness is recessive, and is found on the X chromosome. If a woman has this recessive allele it will have no effect, if she has a dominant allele on her other X chromosome that will mask the recessive one. She will only be colour blind if she has two recessive alleles. If a man carries the recessive allele, the situation is different. There is no corresponding allele on the Y chromosone so there can be no dominant allele to mask the effects of the recessive one. As a result, men are much more likely to be colour blind than women. The same applies to the gene for haemophilia.

A woman can have one allele for colour blindness, or for haemophilia, and yet appear normal. A woman who has the allele for a disorder but shows no symptoms of it is called a carrier. In the case of colour blindness this is not so serious, but haemophilia is a serious disease, so it is important to know if a woman is a carrier. If a man married a woman who was a haemophilia carrier, then their children would inherit either a normal X chromosome or a Y chromosome from their father, and a normal X chromosome or a haemophilia-carrying X chromosome from their mother. So on average there would be a 50 per cent chance that any son born would be a haemophiliac, and a 50 per cent chance that any daughter would be a carrier of the disease.

Different sex genes

Male and female **organisms** are not all distinguished genetically in the same way as in mammals. In some species, for example, it is females that have the Y chromosome. In honey bees, the males (the drones) are **haploid** – they have only a single set of chromosomes rather than **homologous pairs**. The males develop from unfertilized eggs, while the female eggs are fertilized by the queen bee.

Variety – the spice of life

There is a huge range of dog breeds, from tiny toy dogs like the Pekinese to tall, long-legged hunting dogs like the Irish wolfhound, but all dogs belong to the same **species** – one dog can interbreed with another. The differences in dog **phenotypes** that we see are called variation in that species. We can see variation in other species, including humans, although perhaps it is not so great as in dogs.

Variations that can be passed from one generation to the next are called inherited characteristics. They are **genetic** in nature. Other variations are due to environmental factors. One person might sit eating in front of the TV all the time and not take much exercise. Another might play sports, eat well and develop muscles. The muscle development is called an acquired characteristic. Genes alone cannot give you big muscles, and neither can you pass on big muscles to your offspring. Many variations we see in **organisms** are a result of genetic and environmental factors working together.

Gene combinations

How does all this variety come about? We have already mentioned **mutations**, mistakes that are made when **DNA** is copied. One example is the appearance of cats with curly-haired coats, called the 'rex' variety, which is thought to have arisen as a result of a mutation.

However, the main reason why one individual varies from another is because each one possesses a different combination of genes. Part of the reason for this is as a result of what happens when **gametes** are formed. It is pure chance which member of a **homologous pair** of **chromosomes** will find itself in a particular gamete. This is called free assortment.

Examples of the wide variation in colour, size and other characteristics between different breeds of dog.

Genes that are on the same chromosome are not separate in the same way as genes on different chromosomes. However, in one stage of the process of **meiosis**, there is a chance for genes from the homologous chromosomes to mix and match.

Crossing over

In the first stage of meiosis I (**prophase** I), we saw that pairs of homologous chromosomes join together to form double chromosomes, each with two **chromatids**. At this point a process called crossing over takes place. In crossing over, each chromatid can break somewhere along its length. When this happens, the broken pieces can join up again as before, or it may swap over with the same piece in its homologous partner – the chromatids swap genes. When the chromatids eventually separate to form gametes, there are new combinations of genes on each chromatid. This process of crossing over is unique to meiosis.

Let us look at an example. Before crossing over, the joined homologous chromosomes consist of four chromatids: two copies of each homologous chromosome. During crossing over, a section B from one chromatid swaps with a section b on its partner. On the other pair of chromatids, a section C on one chromatid swaps with a section c on its partner. So after crossing over all four chromatids are different.

The process of crossing over creates a much greater range of variation between gametes. Which particular male and female gametes join to form a fertilized **egg**, and therefore which set of genes is inherited from each parent, is random. We see this genetic variation reflected in the physical variety of individuals.

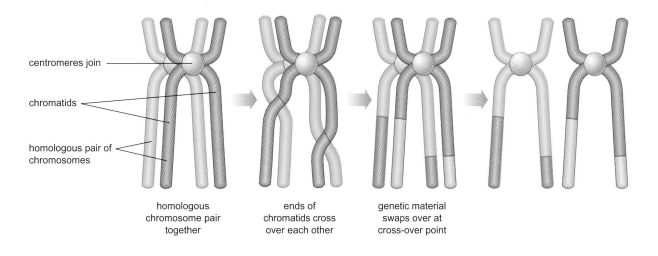

centromeres join

chromatids

homologous pair of chromosomes

homologous chromosome pair together

ends of chromatids cross over each other

genetic material swaps over at cross-over point

An illustration of crossing-over during meiosis.

Genes and the environment

The appearance of an **organism** is not only the result of its genes. The place where an organism lives, and the events in its life, also have a great effect. The influence of genes and the influence of the environment are inextricably linked. The two interact with each other and are difficult to separate.

Effects of the environment

We have already looked at one effect of environment – how the amount of exercise someone takes affects their muscle development. A more important influence in many **species** is the availability of food. Animals that have a limited food supply, especially when they are young, are generally smaller than animals that have plenty of food. However, these characteristics are not passed on to the animal or plant's offspring, because they are not coded into the genes.

The environment can have other effects on how an organism develops. In some cases, its environment determines the sex of an animal. The sex of a crocodile or a turtle, for instance, depends on the temperature of the **egg** while it is incubating. Even more remarkably, fish called blue-headed wrasse have only one male in each shoal. If the male dies, one of the females in the shoal will change sex and become a male.

Genes and evolution

We have seen that the variation between individuals of a species is a result of **mutations** and of different combinations of genes. This variation has an important purpose. Within a group of organisms of the same species living in the same area (a population), some combinations of genes will give rise to characteristics that give an advantage to a particular individual in the struggle for survival. Those individuals best equipped to survive in their environment are the ones that are most likely to have offspring.

Darwin's drawing of the beaks of different Galapagos finches. These birds were early colonizers of the islands, and soon several species arose, adapted to eating different types of food.

Over millions of years, genetic changes in the ancestors of these dolphins and other marine mammals enabled them to adapt to life in the seas.

This filtering out of the less well-adapted organisms is known as natural selection. Genes with a positive effect are selected in favour of genes with a less good effect. Natural selection makes it possible for a population to adapt and survive if its environment changes, or if it has to move to a new environment. As an example, let's look at the way that modern giraffes are thought to have evolved their long necks.

The ancestors of today's giraffes had necks no longer than other similar animals, but food must have become scarcer, for some reason – perhaps the population of giraffe ancestors grew too large, or perhaps other types of animal began to compete with them for food Another possibility is that the climate began to change, and the giraffes' food did not grow so well in the new conditions.

Genetic variation within the giraffe population produced some individuals that had slightly longer necks. These taller giraffes were able to reach food that the other giraffes could not, and so more of them survived and produced offspring.

Natural selection and genetic variation have been the driving force behind the evolution of living things since the first cells appeared. Together they have produced the enormous diversity of life that we see today.

The genetic code

The table below shows the 64 possible ways in which the four **bases** in **RNA** can be combined in groups of three (triplets), and the **amino acid** that each of these triplets codes for. Note that the code has the base uracil (U) instead of thymine (T), because uracil replaces thymine in RNA.

Suppose you want to know the genetic code for methione, the 'start' signal amino acid. Find methionine in the table, then look along the row to the left to the first column to find the first base of the three-base genetic code. You will see that the first base is adenine (A). Now look at the top of the column methione is in, to find the second base of the code: uracil (U). Finally, look along the methionine row to the right to the last column to find the third base in the code: guanine (G). So the start code is AUG.

First base	Second base				Third base
	Uracil (U)	Cytosine (C)	Adenine (A)	Guanine (G)	
Uracil (U)	Phenylalanine	Serine	Tyrosine	Cysteine	Uracil (U)
	Phenylalanine	Serine	Tyrosine	Cysteine	Cytosine (C)
	Leucine	Serine	STOP	STOP	Adenine (A)
	Leucine	Serine	STOP	Tryptophan	Guanine (G)
Cytosine (C)	Leucine	Proline	Histidine	Arginine	Uracil (U)
	Leucine	Proline	Histidine	Arginine	Cytosine (C)
	Leucine	Proline	Glutamine	Arginine	Adenine (A)
	Leucine	Proline	Glutamine	Arginine	Guanine (G)
Adenine (A)	Isoleucine	Threonine	Asparagine	Serine	Uracil (U)
	Isoleucine	Threonine	Asparagine	Serine	Cytosine (C)
	Isoleucine	Threonine	Lysine	Arginine	Adenine (A)
	Methionine (START)	Threonine	Lysine	Arginine	Guanine (G)
Guanine (G)	Valine	Alanine	Aspartic acid	Glycine	Uracil (U)
	Valine	Alanine	Aspartic acid	Glycine	Cytosine (C)
	Valine	Alanine	Glutamic acid	Glycine	Adenine (A)
	Valine	Alanine	Glutamic acid	Glycine	Guanine (G)

Glossary

alleles two versions of a gene, coding for the same characteristic, found on homologous chromosomes

amino acids a range of small molecules that form the subunits from which proteins are built. About 20 amino acids are used in living cells.

anaphase the third phase of mitosis and meiosis, in which chromosomes move to opposite ends of the cell

angiosperm a flowering plant or tree

base a type of chemical found in genetic material. There are four different bases in DNA.

blastula a hollow, fluid-filled ball of cells, an early stage in the growth of a fertilized egg

cell cycle the process of growth and division that occurs again and again in a cell

centrioles structures that appear in a cell during mitosis, which anchor the ends of the spindle

centromere the region where the two chromatids of a chromosome are joined

chromatids the two identical copies of a chromosome that are seen during mitosis and meiosis

chromatin a dark area sometimes seen in a cell nucleus. It is the material from which chromosomes are made – DNA and proteins.

chromosomes structures in the cell nucleus, which contain the cell's DNA or genetic material

cotyledons the seed leaves of a flowering plant, which supply the young plant with food until its first true leaves grow

cytokinesis final stage of cell division, in which the cell splits into two parts

cytoplasm all contents of a cell, outside the nucleus

diploid organism, tissue, or a cell in which there are two versions of each chromosome. Most cells are diploid.

DNA (deoxyribonucleic acid) the genetic material – a substance that carries instructions for constructing, maintaining and reproducing a living cell

dormancy in difficult conditions, an organism may go into a resting state, called dormancy. This conserves the organism's energy for a time when conditions improve.

ectoderm the outer layer of cells in a blastula, from which the skin and nervous system develop

egg the female gamete or sex cell, one of the two cells that join together during sexual reproduction to form a new organism

embryo a young organism in its earliest stages of growth

endoderm the inner layer of cells in a blastula, from which the lining of the gut and other organs develop

endosperm a layer within a seed which acts as a food store for the plant embryo

enzymes special proteins that control the speed of chemical reactions within the cell. Each reaction has its own specific enzyme.

eukaryote one of the two basic types of cell. All kinds of living things except bacteria are made up of eukaryote cells. They contain a nucleus.

gametes the sex cells in organisms that reproduce sexually. During sexual reproduction a male and a female gamete fuse, or join, to form a new organism.

gastrula a stage in the development of an animal that begins with a production of the germ layers

genetic material material (such as DNA) to do with genes, the units that pass on characteristics from parent to offspring

genetics the study of heredity, or how characteristics are passed on from one generation of living things to the next

genotype the genetic make-up of an individual, the sum total of all its alleles

germination the start of growth in a young plant, when the seed splits open and the roots and shoots begin to grow

haploid a cell containing only one copy of each chromosome such as gametes and most fungal cells

heterozygous an organism is heterozygous for a characteristic if it has two different versions of the gene (allele) for that characteristic

homologous pair a pair of chromosomes that are similar in shape and size, and carry genes for the same set of characteristics

homozygous an organism is homozygous for a characteristic if it has two identical versions of the gene for that characteristic

immune system the system within humans and other animals that helps defend against disease

interphase the stage in the cell cycle between divisions, when the cell is growing and copying its DNA

meiosis the type of nuclear division that occurs during gamete formation, in which the number of chromosomes in the nucleus is halved

meristem plants only grow in certain areas, such as the tips of roots and shoots. These growing areas are called meristems.

mesoderm the middle layer of cells in a bastula, from which the muscles, bones and other tissues develop

messenger RNA (mRNA) the molecule that carries the genetic code for a protein out of the nucleus to the ribosomes, where protein assembly takes place

metabolic rate the speed of many different chemical reactions that happen in cells, taken as a whole

metaphase the second of mitosis stage or meiosis, in which the chromosomes move to the centre of the cell prior to being separated

microtubules tiny, hollow protein tubes, which have the ability to shorten or lengthen. In mitosis, the microtubules form the spindle that pulls the chromosomes apart.

mitochondria organelles within eukaryote cells that are responsible for energy release

mitosis the type of nuclear division in which the number of chromosomes stays the same, such as asexual reproduction, growth, repair, etc

mutation any change in a cell's genes or in a chromosome's structure, or a change in the number of chromosomes in a cell

nucleolus a darker area within the nucleus, where genes are being used as templates to make RNA for ribosomes

nucleotides the small subunits from which DNA and RNA are made. Nucleotides consist of a sugar molecule, a phosphate group and a base.

nucleus the biggest and most obvious structure inside a eukaryote cell. The nucleus is where the cell's genetic material is stored.

organelles membrane-bound structures within a cell that are responsible for specific jobs within that cell

organism any type of living thing

phenotype the outward appearance of an organism, and the characteristics that result from its genetic make-up

photosynthesis process by which green plants and some other organisms use the energy of sunlight to assemble glucose from carbon dioxide and water

pollination part of sexual reproduction in flowering plants, in which pollen grains (containing male gametes) from a flower are carried to the stigma (part of the female reproductive structure) of the same or another flower

prokaryotes one of the two basic cell types. Bacteria are the only prokaryotes. Their cells are simpler than those of all other organisms and do not have a nucleus.

prophase the first stage of mitosis or meiosis, in which the chromosomes first become visible through a microscope. In prophase of meiosis I homologous chromosmes pair and exchange genetic material between them.

proteins substances which make up many cell structures and control its reactions. Proteins are large molecules made up of many subunits called amino acids.

protistans a large, loose grouping of organisms, many of them single-celled

ribosomes organelles within eukaryote cells that are responsible for protein synthesis

RNA (ribonucleic acid) a substance related to DNA which carries out a number of important tasks in protein synthesis

species the basic unit by which organisms are grouped. Organisms of the same species can breed together and produce fertile offspring.

sperm the male gamete or sex cell, one of the two cells that join together during sexual reproduction to form a new organism

spindle in mitosis and meiosis, a structure made from microtubules that attaches to the chromosomes and pulls the two copies of each chromosome apart

stigma part of the female reproductive structure in flowering plants, the site to which pollen grains attach

telomere a structure on the end of a chromosome that protects it from damage during cell division. Each time a cell divides, part of the telomeres is lost.

telophase the fourth and final phase of mitosis or meiosis, in which new nuclei are formed

transcription the production of a single-stranded RNA molecule from a section of DNA

transfer RNA (tRNA) a type of RNA that carries amino acids to the ribosomes and is involved in attaching them to make proteins

vesicles small, round cavities in the cell, surrounded by a membrane, which carry materials around the cell or out of it

zygote a fertilised egg cell, the very first cell of a new organism

Further reading and websites

Books

Coordinated Science: Higher Biology, Richard Fosbery and Jean McLean,
     Heinemann Library, 1996
Improving Nature? The Science and Ethics of Genetic Engineering,
     Michael J Reiss, Roger Straughan, Cambridge University Press, 2001
Life processes: Cells and Systems, Holly Wallace, Heinemann Library, 2002
Microlife Series: four books about the world of micro-organisms
     A World of Micro-organisms, Robert Snedden, Heinemann Library, 2001
     Scientists and Discoveries, Robert Snedden, Heinemann Library, 2001
     The Benefits of Bacteria, Robert Snedden, Heinemann Library, 2001
     Fighting Infectious Disease, Robert Snedden, Heinemann Library, 2001
Science at the Edge, Cloning, Sally Morgan, Heinemann Library, 2002
Science at the Edge, Genetic Modification of Food, Sally Morgan,
     Heinemann Library, 2002
Science Fact Files: Genetics, Richard Beatty, Hodder Wayland, 2001

Websites

Cells Alive (http://www.cellsalive.com/)
     A good introduction to the cell, with homework help.
Bugs in the News (http://people.ku.edu/~jbrown/bugs.html)
     Lots of answers for those curious about cells and other very small things.
How stuff works – Cells (http://www.howstuffworks.com/cell.htm)
     Worth a visit to discover unusual aspects of life at cell level.
Cells R Us (http://www.icnet.uk/kids/cellsrus/cellsrus.html)
     Animated cell division.
Cell biology tutorial (http://www.biology.arizona.edu/cell_bio/cell_bio.html)
     Revision on the structure and function of the cell.
Natural history of genes (http://gslc.genetics.utah.edu/)
     Challenge nature and build your own DNA molecule.
Frank Potter's Science gems (http://sciencegems.com/life.html)
     A wealth of references all about cells.
Slouching towards creation –
     http://www.pathfinder.com/TIME/cloning/home.html
     Scientific breakthroughs in cloning.
I can do that! (http://www.eurekascience.com/IcanDoThat/links.htm)
     Cells, DNA and all that brought to life.

Index